The True Story of
Story of
Pocahontas

風中奇緣

商務印書館

出版説明

　　本館一向倡導優質閱讀，近年來連續推出了以"Q"為標識的
"Quality English Learning 優質英語學習"系列，其中《讀名著學英語》叢
書，更是香港書展入選好書，讀者反響令人鼓舞。推動社會閱讀風氣，推
動英語經典閱讀，藉閱讀拓廣世界視野，提高英語水平，已經成為一種潮
流。

　　然良好閱讀習慣的養成非一日之功，大多數初、中級程度的讀者，常
視直接閱讀厚重的原著為畏途。如何給年輕的讀者提供切實的指引和幫
助，如何既提供優質的學習素材，又提供名師的教學方法，是當下社會關
注的重要問題。針對這種情況，本館特別延請香港名校名師，根據多年豐
富的教學經驗，精選海外適合初、中級英語程度讀者的優質經典讀物，有
系統地出版了這套叢書，名為《Black Cat 優質英語階梯閱讀》。

　　《Black Cat 優質英語階梯閱讀》體現了香港名校名師堅持經典學習的
教學理念，以及多年行之有效的學習方法。既有經過改寫和縮寫的經典名
著，又有富創意的現代作品；既有精心設計的聽、説、讀、寫綜合練習，
又有豐富的歷史文化知識；既有彩色插圖、繪圖和照片，又有英美專業演
員朗讀作品的 CD。適合口味不同的讀者享受閱讀之樂，欣賞經典之美。

　　《Black Cat 優質英語階梯閱讀》由淺入深，逐階提升，好像參與一個
尋寶遊戲，入門並不難，但要真正尋得寶藏，需要投入，更需要堅持。只
有置身其中的人，才能體味純正英語的魅力，領略得到真寶的快樂。當英
語閱讀成為自己生活的一部分，英語水平的提高自然水到渠成。

商務印書館（香港）有限公司
編輯部

使用說明 _____

 應該怎樣選書?

按閱讀興趣選書

《Black Cat 優質英語階梯閱讀》精選世界經典作品,也包括富於創意的現代作品;既有膾炙人口的小說、戲劇,又有非小說類的文化知識讀物,品種豐富,內容多樣,適合口味不同的讀者挑選自己感興趣的書,享受閱讀的樂趣。

按英語程度選書

《Black Cat 優質英語階梯閱讀》現設 Level 1 至 Level 6,由淺入深,涵蓋初、中級英語程度。讀物分級採用了國際上通用的劃分標準,主要以詞彙(vocabulary)和結構(structures)劃分。

Level 1 至 Level 3 出現的詞彙較淺顯,相對深的核心詞彙均配上中文解釋,節省讀者查找詞典的時間,以專心理解正文內容。在註釋的幫助下,讀者若能流暢地閱讀正文內容,就不用擔心這一本書程度過深。

Level 1 至 Level 3 出現的動詞時態形式和句子結構比較簡單。動詞時態形式以現在時(present simple)、現在時進行式(present continuous)、過去時(past simple)為主,句子結構大部分是簡單句(simple sentences)。此外,還包括比較級和最高級(comparative and superlative forms)、可數和不可數名詞(countable and uncountable nouns)以及冠詞(articles)等語法知識點。

Level 4 至 Level 6 出現的動詞時態形式,以現在完成時(present perfect)、現在完成時進行式(present perfect continuous)、過去完成時(past perfect continuous)為主,句子結構大部分是複合句(compound sentences)、條件從句(1st and 2nd conditional sentences)等。此外,還包括情態動詞(modal verbs)、被動形式(passive forms)、動名詞(gerunds)、

短語動詞（phrasal verbs）等語法知識點。

　　根據上述的語法範圍，讀者可按自己實際的英語水平，如詞彙量、語法知識、理解能力、閱讀能力等自主選擇，不再受制於學校年級劃分或學歷高低的約束，完全根據個人需要選擇合適的讀物。

② 怎樣提高閱讀效果？

　　閱讀的方法主要有兩種：一是泛讀，二是精讀。兩者各有功能，適當地結合使用，相輔相成，有事半功倍之效。

　　泛讀，指閱讀大量適合自己程度（可稍淺，但不能過深）、不同內容、風格、體裁的讀物，但求明白內容大意，不用花費太多時間鑽研細節，主要作用是多接觸英語，減輕對它的生疏感，鞏固以前所學過的英語，讓腦子在潛意識中吸收詞彙用法、語法結構等。

　　精讀，指小心認真地閱讀內容精彩、組織有條理、遣詞造句又正確的作品，着重點在於理解"準確"及"深入"，欣賞其精彩獨到之處。精讀時，可充分利用書中精心設計的練習，學習掌握有用的英語詞彙和語法知識。精讀後，可再花十分鐘朗讀其中一小段有趣的文字，邊唸邊細心領會文字的結構和意思。

　　《Black Cat 優質英語階梯閱讀》中的作品均值得精讀，如時間有限，不妨嘗試每兩個星期泛讀一本，輔以每星期挑選書中一章精彩的文字精讀。要學好英語，持之以恆地泛讀和精讀英文是最有效的方法。

③ 本系列的練習與測試有何功能？

　　《Black Cat 優質英語階梯閱讀》特別注重練習的設計，為讀者考慮周到，切合實用需求，學習功能強。每章後均配有訓練聽、説、讀、寫四項技能的練習，分量、難度恰到好處。

聽力練習分兩類，一是重聽故事回答問題，二是聆聽主角對話、書信朗讀、或模擬記者訪問後寫出答案，旨在以生活化的練習形式逐步提高聽力。每本書均配有 CD 提供作品朗讀，朗讀者都是專業演員，英國作品由英國演員錄音，美國作品由美國演員錄音，務求增加聆聽的真實感和感染力。多聆聽英式和美式英語兩種發音，可讓讀者熟悉二者的差異，逐漸培養分辨英美發音的能力，提高聆聽理解的準確度。此外，模仿錄音朗讀故事或模仿主人翁在戲劇中的對白，都是訓練口語能力的好方法。

閱讀理解練習形式多樣化，有縱橫字謎、配對、填空、字句重組等等，注重訓練讀者的理解、推敲和聯想等多種閱讀技能。

寫作練習尤具新意，教讀者使用網式圖示（spidergrams）記錄重點，採用問答、書信、電報、記者採訪等多樣化形式，鼓勵讀者動手寫作。

書後更設有升級測試（Exit Test）及答案，供讀者檢查學習效果。充分利用書中的練習和測試，可全面提升聽、說、讀、寫四項技能。

④ 本系列還能提供甚麼幫助？

《Black Cat 優質英語階梯閱讀》提倡豐富多元的現代閱讀，巧用書中提供的資訊，有助於提升英語理解力，擴闊視野。

每本書都設有專章介紹相關的歷史文化知識，經典名著更有作者生平、社會背景等資訊。書內富有表現力的彩色插圖、繪圖和照片，使閱讀充滿趣味，部分加上如何解讀古典名畫的指導，增長見識。有的書還提供一些與主題相關的網址，比如關於不同國家的節慶源流的網址，讓讀者多利用網上資源增進知識。

Contents

Special Features:

The story is recorded in full. 故事錄音

 These symbols indicate the beginning and end of the extracts linked to the listening activities.   聽力練習開始和結束的標記

Is this a true story?

Yes, this is a true story. The Indian princess Pocahontas was born near Jamestown in 1595. Captain John Smith wrote many of the events of this story in his books.

Some details were added and other details were forgotten. But the story you will read in this book really happened! This is the **true** story of Pocahontas.

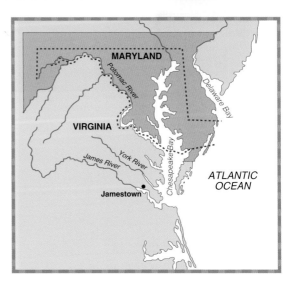

Pocahontas in London (1616) by unknown artist.

This is a map of the Virginia Colony and the settlement of Jamestown on the James River.

John Smith gives a crown to Wahunsonacock in a 19th-century painting.

John Smith and the colonists wanted to keep peaceful relations with Pocahontas' father, Chief Powhatan. Chief Powhatan's other name was Chief Wahunsonacock.

King James I sent presents to the chief. John Smith gave the chief a beautiful crown and proclaimed him King of the Powhatan Empire. However, Chief Wahunsonacock did not know that the Powhatan Empire remained under King James!

Do you know these words?

weapons
武器

corn
玉米

sailing ship
帆船

beads and trinkets
珠寶

bow and arrow
弓箭

longhouse
長屋

gold and riches
金銀財寶

 Match these definitions with the words in the tepee.

a. A small wooden house is a ....

b. .... travel to another land to live there permanently.

c. A .... is a wild, uncultivated land. There are no people.

d. A small Indian boat is a ....

e. .... are Indian men of magic.

f. A .... cultivates the land.

canoe
farmer
settlers
wilderness
hut or cabin
medicine men

CHAPTER ONE

The Sailing Ship

May was a beautiful month in the land of the Algonquin Indians. There were tall trees and colorful flowers everywhere. The sky and the sea were deep blue.

Pocahontas was the favorite [1] daughter of Chief Powhatan. She was an Indian princess. Chief Powhatan was a powerful chief of the Algonquin tribe.

Pocahontas was eleven years old. She was a lovely young girl with black hair and dark eyes. She wore a buckskin dress [2] and moccasins [3]. She had a feather [4] in her hair.

---

1. **favorite** : 最喜愛的。

2. **buckskin dress** :

3. **moccasins** :

4. **feather** :

Pocahontas

Pocahontas was always happy. She ran in the forest and danced in the fields. She sat on the hill and looked at the blue sea.

On May 6, 1607, Pocahontas sat on the hill and she saw something strange in the bay. It was a big sailing ship! She was very surprised. The sailing ship was something new. It had big white sails and flags. She looked at it for a long time. Where did it come from? Why was it there? She was very excited.

Pocahontas ran to her village to tell her father and brother the news.

"Father, father!" she said. "There's a big sailing ship in the bay! It has white sails and colored flags!"

Her father, Chief Powhatan, was outside his longhouse. He was a tall man, with long black hair. He wore the feathers of an Indian chief on his head. He listened to the news, but he was not happy.

"The white men are here," Powhatan said sadly. "This is bad news for our people. There is no peace with the white men here."

Nantaquas was Pocahontas' brother. He was eighteen years old. He was a strong Indian warrior[1]. He looked at his sister and said, "In the past the white people killed the Indians. They want to take our land."

"Oh, Nantaquas, let's go to see the white people. Let's go to see their big ship!" said Pocahontas.

"You can go to see them, but don't go near the ship. Stay far away! Be careful!" said Powhatan. "Remember, Pocahontas, white men are dangerous."

Nantaquas and Pocahontas went to the river. There were many canoes near the river. They got into a small canoe. Then

1. warrior : 武士。

Pocahontas

Nantaquas paddled [1] down the river to the bay. In the bay they saw the big sailing ship. There was a tall, white man on the ship. He smiled at them. There were other white men too. They all looked at the small canoe and at the two Indians.

Pocahontas smiled at the white man. "Let's go to the ship," she said.

"No," said Nantaquas. "It's dangerous. We don't know who these white men are. We can look and then go home. We must obey our father."

The white man on the ship smiled again. "Look, Nantaquas," said Pocahontas, "that man has red hair and white skin! He's smiling at us. Look at his clothes! They are strange."

Nantaquas turned the canoe and paddled up the river. When they arrived home, Nantaquas said, "I saw a big sailing ship. There were many white men on the ship." Powhatan talked to his medicine men[2] and tribe advisers [3]. They all sat inside the longhouse for a long time.

He said, "White men bring us problems. They have a strange magic. They carry thundersticks [4] to kill our people. They want to take our land, our lakes, our rivers and our forests. They can stay for a short time but they cannot stay here forever. Let's watch them and see what they do!"

1. **paddled**：划船。
2. **medicine men**：醫師。
3. **tribe advisers**：部落長老。
4. **thundersticks**：火槍（印第安人用語）。

UNDERSTANDING THE TEXT

1 **Are sentences 1-7 "Right" (A) or "Wrong" (B)?**
If there is not enough information to answer "Right" (A) or "Wrong" (B),
choose "Doesn't say" (C).

0. The story begins in the month of May.
(A) Right   **B** Wrong   **C** Doesn't say

1. Chief Powhatan was the chief of the Nantaquas tribe.
A Right   **B** Wrong   **C** Doesn't say

2. Pocahontas was the granddaughter of an Indian chief.
A Right   **B** Wrong   **C** Doesn't say

3. She sang many Indian songs.
A Right   **B** Wrong   **C** Doesn't say

4. On May 6, 1607, Pocahontas saw two sailing ships in the bay.
A Right   **B** Wrong   **C** Doesn't say

5. Chief Powhatan had five children.
A Right   **B** Wrong   **C** Doesn't say

6. Pocahontas and her brother went to see the sailing ship.
A Right   **B** Wrong   **C** Doesn't say

7. A white man on the ship smiled at Pocahontas and her brother.
A Right   **B** Wrong   **C** Doesn't say

2 **Look at the adjectives** （形容詞）**below. Then look at the nouns** （名詞）
in the box. Put each noun next to the adjective that best describes it.
One is done for you. There can be more than one noun for each
adjective.

> man   dress   food   eyes   child   biscuit
> weather   tree   sky   daughter   flowers

tall: _tree,_ ....

colorful: ....

sweet: ....

blue: ....

bad: ....

strong: ....

15

3 **Complete the five conversations.**
For questions 1-5, mark A, B or C.

0. Do you like corn?

A ☐ Yes, I have.
B ☑ Yes, I do.
C ☐ No, I can't.

1. Can I go to see the ship?

A ☐ At two o'clock.
B ☐ No, you don't.
C ☐ Yes, you can.

2. When will Nantaquas come back?

A ☐ I don't know.
B ☐ He will.
C ☐ Yesterday.

3. Where's the big canoe?

A ☐ It's small.
B ☐ No, it isn't.
C ☐ Near the river.

4. Can you give me the paddles?

A ☐ Yes, I do.
B ☐ Yes, of course.
C ☐ No, I'm not.

5. Do you want to play in the forest?

A ☐ Yes, I played.
B ☐ Tomorrow morning.
C ☐ Yes, please.

4 **Substitute the number with the correct word in the arrow.**

eight   three hundred   eleven   fourteen
thirty-two   eighteen   seventy-five

a. Pocahontas is 11 .... years old.

b. Her brother Nantaquas is 18 .... years old.

c. There are 300 .... Indians in the forest.

d. Chief Powhatan has 8 .... brothers.

e. Pocahontas saw 14 .... red flowers on the hill.

f. There were 32 .... trees near the sea.

g. There were 75 .... men on the sailing ship.

5 **Listen to this dictation carefully. Then listen to it again and fill in the gaps.**

Pocahontas was an Indian [1] .... She was always

[2] .... She lived near the [3] .... She was [4] ....

years old. She had [5] .... hair. Her [6] .... was Chief

Powhatan. Her [7] .... was called Nantaquas. He was a strong

warrior. He was eighteen years [8] .... Chief Powhatan didn't

[9] .... white men.

17

CHAPTER TWO

Pocahontas meets John Smith

Captain John Smith and his men were happy to be in Chesapeake Bay. They wanted to establish [1] a small settlement [2] there. Captain Smith called the big river the James River, after King James I of Britain. On May 13, 1607, he established the small settlement called Jamestown. Jamestown was on the James River. In Jamestown the settlers built some huts [3], a storehouse [4] and a church.

There were about 100 men in Jamestown in 1607. Most of them were English gentlemen. They came to the New World to find gold and riches. They did not want to be farmers. John Smith was angry with them. He said, "You must all plant crops [5],

---

1. **establish** : 建立。
2. **settlement**: 居民點。
3. **huts** :
4. **storehouse** : 倉庫。
5. **plant crops** : 種植糧食。

Pocahontas meets John Smith

hunt and fish. You must not be lazy!"

In Jamestown there was little food. One day Captain Smith and his men went into the forest to look for food. They walked for a long time. Then they met a big group of Indians.

The Indians attacked them with their bows and arrows. They killed one of Smith's men. John Smith and his men killed two Indians. Then the Indians captured [1] John Smith and took him away.

After a long walk, John Smith stood in front of Chief Powhatan and his tribe. Everyone was silent. Pocahontas stood next to her father. She looked at John Smith. He was very tall. She looked at his red hair, his blue eyes and his white skin. He was very different from the Indian men.

John Smith spoke to the Indians in sign language [2] and a few Indian words.

"Great chief, I am a friend. My men and I want to live in peace with you."

Powhatan and his medicine men did not like him.

John Smith gave a compass [3] to the great chief.

---

1. **capture**：捕獲。
2. **sign language**：手語。
3. **compass**： 指南針。

Pocahontas

Powhatan looked at it. He turned it around in his hand. Why did the needle [1] always point in the same direction? He tried to touch the needle but a piece of ice was in front of it! The ice wasn't cold. It didn't melt! Powhatan thought it was magic. All the Indians of the tribe looked at the compass. They were surprised at the white man's magic.

Pocahontas liked John Smith and his magic, but her father didn't like him. That afternoon John Smith and his men killed two Indians. Chief Powhatan and his tribe were very angry. Now John Smith must die!

Two Indian warriors pushed Captain Smith to the ground. They put his head on a very big stone. Then the Indians picked up [2] another big stone. They wanted to kill John Smith! When Pocahontas saw this, she said, "No, father. Please don't kill him. He isn't a bad man." Powhatan said, "No! He and his men killed two Indians. He must die."

The two Indians were ready to kill Captain Smith. One Indian raised his hand.

"No!" said Pocahontas. She jumped forward and put her head above Captain Smith's head. "Please father, he must not die! Save him!" said Pocahontas.

Powhatan looked at his favorite daughter. He immediately told the two Indians to stop. Everyone was surprised at Pocahontas' courage. Pocahontas saved John Smith's life.

1. **needle**：指針。
2. **pick up**：舉起。

Pocahontas

After this Pocahontas and John Smith became great friends. John Smith taught her English and she taught him the Indian language. He gave her beautiful beads and trinkets. He told her about London and its enormous [1] buildings. Pocahontas listened to Smith's stories.

"The King of England is called King James I. He lives in a beautiful palace in London," said John Smith.

"Is he your chief?" asked Pocahontas.

"Yes, he's our leader," said John Smith.

"What do the English ladies wear?" asked Pocahontas.

"They wear long, colorful dresses, shoes and hats. They also wear jewels."

"Are the English ladies beautiful?" Pocahontas asked.

"Some are beautiful and some aren't!" said John Smith.

Pocahontas laughed and listened. She dreamed about London.

1. **enormous** : 巨大的。

1 **Choose the correct answer.**

a. In 1607 Captain John Smith established a small settlement called
- [ ] Chesapeake Bay.
- [ ] Jamestown.
- [ ] James River.

b. The new settlers were
- [ ] farmers.
- [ ] hunters.
- [ ] English gentlemen.

c. They wanted to
- [ ] find gold and riches.
- [ ] make maps.
- [ ] plant crops.

d. One day Captain Smith and his men
- [ ] went to visit an Indian tribe.
- [ ] went into the forest to look for food.
- [ ] built a small school.

e. Chief Powhatan and his medicine men decided
- [ ] to be friends with John Smith.
- [ ] to visit Jamestown.
- [ ] to kill John Smith.

f. Pocahontas jumped forward and
- [ ] ran away.
- [ ] put her head above Captain Smith's.
- [ ] pushed away two Indian warriors.

g. Pocahontas saved John Smith's life and
- [ ] they became great friends.
- [ ] John Smith returned to London.
- [ ] Pocahontas went to London.

2 **Read the descriptions below.**
What is the word for each one?
The first letter is already there.
There is one space for each other letter in the word.

0. A small house.

   h u t

1. A place where food and other things are kept.

   s _ _ _ _ _ _ _ _

2. The needle always shows North.

   c _ _ _ _ _ _

3. The leader of a tribe.

   c _ _ _ _

4. Captain Smith established it and called it Jamestown.

   s _ _ _ _ _ _ _ _ _

5. Where King James I lives.

   p _ _ _ _ _

3 **LOOKING AT PICTURES**
Look at the picture on page 23 and answer these questions.

a. Where are Pocahontas and John Smith?

   ....

b. What's the weather like?

   ....

c. What is John Smith doing?

   ....

d. What do you think Pocahontas is saying?

   ....

e. Why are there books on the grass?

   ....

Who did what?
Read the clues and write the name of the person who did the following actions. Some names can be used more than once.

Who...

a. ☐ looked carefully at the compass?

b. ☐ was ready to kill Captain Smith?

c. ☐ built some huts, a storehouse and a church?

d. ☐ established Jamestown?

e. ☐ liked John Smith and his magic?

f. ☐ wanted to find gold and riches?

g. ☐ was angry with the white settlers?

h. ☐ saved John Smith's life?

i. ☐ gave Pocahontas beads and trinkets?

j. ☐ told Pocahontas about London?

1. two Indian warriors
2. settlers
3. Captain John Smith
4. Pocahontas
5. Chief Powhatan

T: GRADE 3

 Topic – School
"John Smith taught Pocahontas English and she taught him the Indian language."

Think about the following questions.

a. Do you like studying English? Why or why not?

b. Which places would you like to visit in Great Britain? Why?

c. Would you like to study English in Great Britain? In which town or city?

26

6 Which notice (A-H) says this (1-5)?
For questions 1-5, mark the correct letter A-H.

0. We won't be long.
1. There will be an excursion next week.
2. We make bread.
3. Your dog can't enter.
4. We don't open in the afternoon.
5. You can eat lunch here.

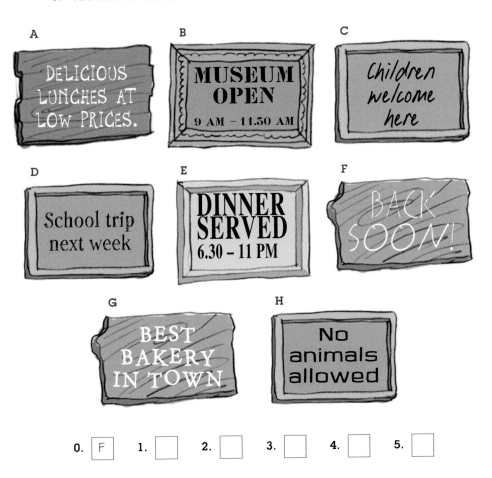

A

DELICIOUS
LUNCHES AT
LOW PRICES.

B

MUSEUM
OPEN

9 AM – 11.50 AM

C

Children
welcome
here

D

School trip
next week

E

DINNER
SERVED
6.30 – 11 PM

F

BACK
SOON!

G

BEST
BAKERY
IN TOWN

H

No
animals
allowed

0. [F]   1. [ ]   2. [ ]   3. [ ]   4. [ ]   5. [ ]

7 Complete this letter.
Write **ONE** word for each space (1-10).

Dear King James,

We arrived in Chesapeake Bay 0 .....on............. May

13, 1607. Our journey 1 .... very difficult.

Our settlement is 2 .... Jamestown and it is

on the James River.

The English gentlemen do 3 .... want to

plant crops. They 4 .... to find gold, but we

must 5 .... crops before winter.

Last week we built 6 .... huts. The Indians

here 7 .... very friendly and they help

8 .... I met 9 .... Indian

girl and 10 .... name is Pocahontas.

From,

Captain John Smith

The Origins[1] of Virginia

I n 1606 King James I of Britain established the Virginia Company to explore the New World. It was named after Elizabeth I, the Virgin Queen.

King James wanted his explorers to find gold, riches and food products in the New World. He also hoped to find a new way to go to India.

King James I (1621) by Daniel Mytens.

Captain John Smith was one of the leaders of the Virginia Company. He was a courageous[2] soldier. When he was young, his life was full of adventure.

He and the Austrians fought against the Turks. The Austrians were very pleased with him. They gave him a horse, a scimitar[3], a coat of arms[4] and the title[5] of captain. Then Captain Smith was captured by the Turks. He was a prisoner in Turkey but he escaped. He crossed Russia, sailed on a pirate ship and returned to England in 1604!

1. **Origins**：起源。
2. **courageous**：勇敢的。
3. **scimitar**：短彎刀。

4. **coat of arms**：
5. **title**：頭銜。

Captain John Smith,
a contemporary engraving.

Captain Smith then worked for the Virginia Company. He took the English settlers to the New World. He was a strong, intelligent [1] leader. He established the settlement of Jamestown. He explored the Chesapeake region and made maps. He made friends with many Indian tribes in the region.

The settlers did not find gold and riches in the new colony. But they found the tobacco [2] plant. In the 17th century the tobacco plant grew only in America.

Replicas of the ships that carried the first settlers to Jamestown in 1617.

1. **intelligent**：聰明的。
2. **tobacco**：煙草。

The American Indians smoked the tobacco leaves. They used the leaves as medicine too. The tobacco plant became the most important product of the Virginia Colony. There were many tobacco plantations [1] along the James River. The tobacco leaves were picked [2] and dried. Then they were sent to England and other parts of Europe. This was the beginning of an important business. Soon there were tobacco companies everywhere.

With its tobacco plantations Virginia became a very rich colony. By 1688 it produced about 10 million kilograms of tobacco! Smoking became very popular in America and Europe.

A 1609 advertisement of the Virginia Company. The Company was looking for settlers for the Jamestown colony.

1. **plantations** : 種植園。
2. **picked** : 採摘。

1 **Read the paragraph below. Choose the best word (A, B or C) for each space (1-8).**

The Virginia Company

King James I ⁰..B..... the Virginia Company. He wanted ¹.......... explorers to find riches and food products. When John Smith was young he fought ².......... the Turks and had a lot of adventures. He ³.......... the leader of the Virginia Company. He sailed ⁴.......... the New World and found the tobacco ⁵.......... . The ⁶.......... important product of the colony was tobacco and it ⁷.......... a very important business. By the end of the 17th century, Virginia ⁸.......... about 10 million kilograms of tobacco.

0.	**A** built	**B** established	**C** made
1.	**A** his	**B** it	**C** made
2.	**A** between	**B** against	**C** among
3.	**A** be	**B** is	**C** was
4.	**A** at	**B** to	**C** in
5.	**A** plant	**B** tree	**C** flower
6.	**A** many	**B** much	**C** most
7.	**A** become	**B** became	**C** becoming
8.	**A** produced	**B** made	**C** did

2 **Word puzzle**
Complete the following words.

a. The Virgin Queen
    _ L _ _ _ _ _ _ _

b. Sword used by the Turks
    _ _ _ _ _ T _ _

c. Special pattern or picture which is the sign of a family
    _ _ A _   _ _   _ _ _ _

d. This person is captured and is not free
    _ _ I _ _ _ _ _

e. The settlers did not find this in the new colony
    _ _ L _

CHAPTER THREE

Winter in Jamestown

The hot summer passed and the cool autumn arrived. The Jamestown settlers had little food to eat. Many settlers were ill and weak. They needed help.

When winter arrived there was no food. Pocahontas helped the Jamestown settlers. She asked her father for corn, meat and other food. Pocahontas and other Indians brought the food to Jamestown in big baskets [1]. The courageous Indian princess helped the settlers to live during the cold winter.

Ships came to Jamestown from England. Powhatan was not happy about this. More white men came to the New World. Powhatan was afraid of them. He was afraid of the future.

One winter day Powhatan sent an Indian messenger to Jamestown. He had a message for Captain Smith. "My chief Powhatan wants to speak to you. Follow me." John Smith followed the messenger to Powhatan's village.

1. baskets :

Pocahontas

Powhatan was in his longhouse. John Smith sat next to him. "We have no more food to give to your people. You must all leave this land now," said Powhatan.

"Why must we leave?" asked John Smith.

The two men talked for a long time. At midnight Powhatan said, "It is very late. You can sleep in the small cabin near the river."

Captain Smith accepted the invitation. He went to sleep in the small cabin.

During the night, John Smith heard someone at the door. He got up, opened the door and saw Pocahontas. "What a surprise to see you, Pocahontas! Please come in!"

"Oh, Captain Smith, your life is in danger. My father and the medicine men want to kill you tonight! They don't want white people to stay here. You must run away now."

"Dear princess, you are saving my life again. How can I thank you? What can I give you?" asked Captain Smith.

"Run away *now*! Save yourself!" Pocahontas touched his hand and ran away.

John Smith ran out of the cabin. He walked to Jamestown in the middle of the night. When he arrived in Jamestown he told the settlers that Pocahontas saved his life again. After this adventure, Captain Smith returned to England.

In Pocahontas' village everyone thought that Captain Smith was dead. Everyone said that he was killed by a gunpowder [1] explosion [2].

---

1. **gunpowder**：火藥。
2. **explosion**：爆炸。

KET

 Are sentences 1-8 "Right" (A) or "Wrong" (B)?
If there is not enough information to answer "Right" (A) or "Wrong"
(B), choose "Doesn't say" (C).

0. The Jamestown settlers had lots of food to eat.

A Right **(B)** Wrong **C** Doesn't say

1. They did not want to eat fish.

A Right **B** Wrong **C** Doesn't say

2. Pocahontas brought the settlers food in big baskets.

A Right **B** Wrong **C** Doesn't say

3. An Indian messenger came to Jamestown with a message for Captain Smith.

A Right **B** Wrong **C** Doesn't say

4. Captain Smith did not go to Chief Powhatan.

A Right **B** Wrong **C** Doesn't say

5. In the middle of the night Pocahontas said to Captain Smith, "My father and the medicine men want to kill you."

A Right **B** Wrong **C** Doesn't say

6. Captain Smith walked to Jamestown at night and then returned to England.

A Right **B** Wrong **C** Doesn't say

7. Pocahontas wrote a long letter to Captain Smith.

A Right **B** Wrong **C** Doesn't say

8. Everyone thought that John Smith returned to England.

A Right **B** Wrong **C** Doesn't say

2 Unscramble these words. Then match them with their meanings below.

a. rgesmense ....

b. msmeur ....

c. kaesbt ....

d. dofo ....

e. itwren ....

f. glesnohou ....

1. ☐ something you eat

2. ☐ something used to carry food, etc.

3. ☐ an Indian home

4. ☐ person who brings a message

5. ☐ very cold season

6. ☐ very hot season

BEFORE YOU READ

1 **LOOKING AT PICTURES**
Look at the picture on page 41 and answer these questions.

a. Where is Pocahontas?
....

b. Who are the other women?
....

c. What is Pocahontas doing?
....

d. Do you think she is happy or unhappy?
....

Listen to the first part of Chapter Four and then answer the questions.

1. Who came to Jamestown?
   A ☐ More white settlers.
   B ☐ The Potomac Indians.
   C ☐ King James I.

2. Who declared war on the Virginia Colony?
   A ☐ King James I.
   B ☐ Chief Japazaws.
   C ☐ Chief Powhatan.

3. How old was Pocahontas?
   A ☐ Seventeen years old.
   B ☐ Eleven years old.
   C ☐ Twenty years old.

4. What did Chief Powhatan tell his daughter?
   A ☐ "Don't go to Jamestown."
   B ☐ "You must stay at home."
   C ☐ "All white men are bad."

5. Who was Chief Japazaws?
   A ☐ An explorer from England.
   B ☐ The head of the Potomac tribe.
   C ☐ A messenger.

6. Who was Pocahontas' friend?
   A ☐ Chief Japazaws.
   B ☐ Captain Argall.
   C ☐ Chief Japazaws' wife.

Chief Powhatan declares War!

ore and more white settlers came to Jamestown. Jamestown was part of the Virginia Colony. Chief Powhatan was angry. He declared war on the little colony. There was a lot of fighting.

Pocahontas was 17 years old. Powhatan wanted to protect his favorite daughter. He sent her to live with the Potomac tribe. The Potomac Indians were friends of the white people. Pocahontas was safe with them.

Powhatan said to Pocahontas, "You must stay with the Potomac Indians. You must not go to Jamestown. We are at war with Jamestown."

"Yes father."

Pocahontas liked her life with the Potomacs. Chief Japazaws

was the head of the Potomac tribe. His wife was Pocahontas' friend. Chief Japazaws and his wife were friends of Captain Samuel Argall. Captain Argall was an English explorer. He lived in Jamestown.

One day Captain Argall went to visit Chief Japazaws. When he saw Pocahontas he said to the chief, "Come to see my ship! I want to show you a lot of interesting things. We can eat on the ship."

"You are very kind," said Pocahontas. "I want to see an English ship."

Chief Japazaws and his wife wanted to see the ship too.

Captain Argall took Pocahontas, Chief Japazaws and his wife to the big sailing ship. They looked around the big ship. They saw the tall masts[1] and the white sails. Then they ate delicious food. Pocahontas was very happy and said, "Thank you for a wonderful day, Captain Argall."

At sunset[2] Chief Japazaws and his wife left the ship in a canoe. But Pocahontas did not leave. She was Captain Argall's prisoner! He tricked[3] her!

"Why can't I go with my friends?" asked Pocahontas. She

1. **masts**：船桅，桅杆。

2. **sunset**：

3. **tricked**：欺骗。

Pocahontas

looked at the canoe and saw her friends. Chief Japazaws' wife had a new copper kettle [1] and a basket full of colored beads. Chief Japazaws and his wife helped Captain Argall capture Pocahontas. The copper kettle and the colored beads were their payment [2].

Pocahontas cried. She had no true friends. She asked, "What is happening? Why am I a prisoner?" Captain Argall said, "I don't want to hurt you, Pocahontas. I want to take you to Jamestown and keep you there. When your father returns the weapons he took from us, I can free you. Then you can return home. Your father loves you. He must return the weapons."

Pocahontas was a prisoner but she was not afraid of Captain Argall. She was not afraid of the white people.

Captain Argall took her to Jamestown. Everyone in Jamestown liked Pocahontas. They remembered that she saved Captain Smith's life twice. They also remembered that she brought them food during the long winter.

Everyone in Jamestown was kind and friendly. The women gave her English clothes to wear.

Pocahontas was a beautiful young woman. She learned English manners [3] and customs [4]. She made many friends. Pocahontas became a Christian and her Christian name was Rebecca.

Powhatan did not return the weapons. He sent some corn and some broken weapons. The war continued. Captain Argall was furious because his plan did not work. He kept Pocahontas in Jamestown. She was his prisoner, but she was not unhappy. She liked Jamestown because she learned new things every day.

1. **copper kettle** :
2. **payment** : 報酬。
3. **manners** : 禮儀。
4. **customs** : 習俗。

UNDERSTANDING THE TEXT

 Choose the correct answer.

a. Many settlers came to Jamestown
- ☐ but they soon returned to England.
- ☐ and Chief Powhatan was angry and declared war.
- ☐ and they were expert farmers.

b. The Potomac Indians
- ☐ were at war with the Algonquins.
- ☐ were at war with Jamestown.
- ☐ were friends of the white people.

c. Chief Powhatan sent Pocahontas
- ☐ to England.
- ☐ to live with the Potomac tribe.
- ☐ to Jamestown.

d. Captain Samuel Argall was
- ☐ an English explorer.
- ☐ a friend of Chief Powhatan.
- ☐ Captain John Smith's uncle.

e. He invited Pocahontas, Chief Japazaws and his wife
- ☐ to visit Jamestown.
- ☐ to see his sailing ship.
- ☐ to sail to England with him.

f. Chief Japazaws and his wife
- ☐ helped Pocahontas to escape from Captain Argall.
- ☐ helped Captain Argall to capture Pocahontas.
- ☐ took Pocahontas to Jamestown.

g. In Jamestown Pocahontas was happy
- ☐ but she didn't understand the English language.
- ☐ but one day she escaped.
- ☐ and she made many friends and became a Christian.

2 **Word search**
Find the names of two Indian tribes, two Indian chiefs and two Indians.

M	N	B	D	H	S	H	Y	D	O	A	G	Y
G	A	P	J	C	P	O	W	H	A	T	A	N
X	N	L	E	L	W	U	E	Z	S	M	K	L
O	T	V	G	P	H	T	I	W	G	N	J	R
F	A	L	P	O	C	A	H	O	N	T	A	S
S	Q	U	A	T	N	K	C	F	P	H	P	N
T	U	F	I	O	P	Q	E	S	F	S	A	O
W	A	Y	C	M	V	A	U	G	V	Y	Z	V
I	S	Q	K	A	J	C	X	I	K	R	A	E
B	E	A	L	C	B	F	T	P	N	H	W	R
R	O	C	Z	R	D	G	W	F	L	K	S	N

3 Now fill in the gaps with words from the word search.

a. Chief .... was the leader of the .... tribe.

b. Chief .... was the leader of the .... tribe.

c. .... and .... were brother and sister.

T: GRADE 4

4 **Topic – Food**
"Pocahontas, Chief Japazaws and his wife ate delicious food on the big ship."

Bring in a recipe or a picture of your favorite food or dish.

a. Describe your favorite dish.

b. How often do you have this dish?

c. What are the differences between a typical dish from your town or region and a dish from another region?

43

5 What, when, where, who, why
Fill in the gaps with the correct wh words.

a. .... did Captain Argall give Chief Japazaws and his wife?
He gave them a copper kettle and beads.

b. .... did Captain Argall take Pocahontas?
He took her to Jamestown.

c. .... did Powhatan declare war on Jamestown?
He declared war on Jamestown because many white settlers came to the Virginia Colony.

d. .... did Chief Japazaws and his wife leave the ship?
They left the ship at sunset.

e. .... sent Pocahontas to the Potomac tribe?
Powhatan sent Pocahontas to the Potomac tribe.

6 Nouns（名詞）**are the names of people, places or things. Adjectives** （形容詞）**are words that describe nouns. Adjectives are placed before nouns if they are descriptive**（描述性的）**, and after the verb "to be" if they are predicative**（謂語性的）**. Fill in the gaps with the correct adjectives or nouns from the ones listed below.**

ADJECTIVES	NOUNS
brave	people
dangerous	forest
safe	life
tall	tribe
colored	payment
kind	sails

a. Pocahontas was .... with the Potomac ....

b. They saw the .... masts and white ....

c. The .... beads were their ....

d. The .... of Jamestown were .... and friendly.

e. The .... princess saved John Smith's .... twice.

f. It was .... to walk in the .... at night.

7 When Captain Argall captured Pocahontas, she wrote a short letter to her father. Complete this letter. Use the words in the box to fill in the gaps.

wife   me   kind   prisoner   daughter   settlers
Jamestown   free   true friends   weapons

Dear Father,

I am a ¹..................... in ².....................! Chief Japazaws and his ³..................... tricked me. Captain Argall captured me. I am very sad because I have no ⁴..................... .

The people of Jamestown are ⁵..................... to me but I want to be ⁶..................... . Please return the ⁷..................... you took from the ⁸..................... . Help ⁹....................., father!

Love, your ¹⁰..................... , Pocahontas

8 Have fun with this picture puzzle!

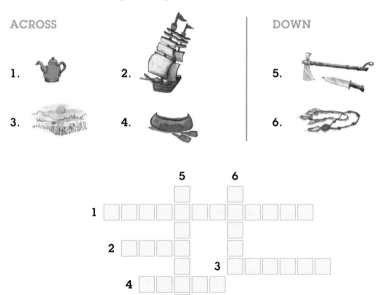

ACROSS

1.   2.

3.   4.

DOWN

5.

6.

Indian Life

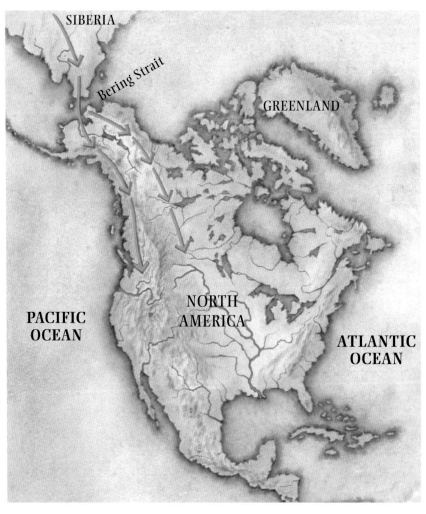

The American Indians came from Asia about 40,000 years ago. They crossed the Bering Strait [1] and went down to North America.

SIBERIA

Bering Strait

GREENLAND

PACIFIC OCEAN

NORTH AMERICA

ATLANTIC OCEAN

---

1. **Bering Strait** : 白令海峽。

These Indians were hunters. They hunted wild animals. After thousands of years, some Indians became farmers. They cultivated[1] corn. Other Indians hunted the buffalo. They used the buffalo for food, clothing and tepees[2].

Buffaloes.

American Indians belonged to different tribes. Every tribe had a chief. There were hundreds of Indian tribes in America.

The American Indians decorated[3] things with beautiful colors. They made masks and decorated them.

1. **cultivated**：種植。
2. **tepees**：圓錐形帳篷。
3. **decorated**：裝飾。

Chief Sitting Bull.

They also made pottery and statues.

Indians painted pictures on big stone walls. These pictures told a story about their life.

Indians were great hunters and courageous warriors. They were strong people. They loved and respected nature and animals.

Indians painted pictures on big stone walls.

Different tribes lived in different kinds of homes: the wigwam [1], the tepee, the longhouse and the pueblo [2]. Indians often decorated their tepees with colorful pictures. The Indians of the Pacific Coast often put big totem poles [3] outside their longhouses.

Indians used these weapons to hunt and fight: bows and arrows, spears [4], knives and tomahawks [5].

Indians often decorated their tepees with colorful pictures.

1. **wigwam** : 棚屋。

2. **pueblo** : 城堡。

3. **totem poles**: 圖騰柱。

4. **spears** :

5. **tomahawks** :

49

When Captain John Smith came to the New World, some important tribes were the Apaches, the Comanches, the Hurons, the Mohicans, the Navajos and the Sioux. At first the Indians were friendly with the white settlers. They taught the settlers how to grow crops. They also taught them how to live in the wilderness.

The white settlers brought the Indians weapons and tools. Soon the settlers wanted to take the Indians' land. Many settlers came to America and they wanted land. The Indians didn't want to lose their land. The Indian wars began. Many people were killed.

Finally in the 1890s the Indian wars ended. The Indians lost their land. America was now the land of the white settlers.

Today American Indians live on reservations [1].

A group of warrior chiefs from the Nez Perce tribe.

1. **reservations** : 印第安人居留地。

1 Are these sentences true (T) or false (F)? Correct the false ones.

	T	F
a. The American Indians came from Eastern Europe about 40,000 years ago.	☐	☐
b. The first Indians hunted wild animals.	☐	☐
c. American Indians lived in small families.	☐	☐
d. The Indians painted pictures on trees.	☐	☐
e. The Indians loved and respected nature and animals.	☐	☐
f. The wigwam and the tepee were Indian homes.	☐	☐
g. The Indians used bows and arrows, spears, knives and tomahawks to hunt and fight.	☐	☐
h. The Apaches, the Comanches and the Mohicans were not important tribes.	☐	☐
i. The settlers took the Indians' land.	☐	☐
j. Today American Indians live on reservations.	☐	☐

2 Odd one out!
Circle the word that doesn't belong to the same category.

a. Indian   English   Spanish   pictures   French
b. Asia   buffalo   America   Europe   Africa
c. tomahawk   Navajo   Sioux   Mohican   Apache
d. tepee   totem poles   pueblo   longhouse   wigwam
e. hunter   warrior   farmer   nature   explorer

3 Now use the odd words to fill in the gaps.

a. Some Indians put big .... outside their longhouses.
b. Indians hunted ....
c. The Indians loved and respected ....
d. The .... was used to hunt and fight.
e. Indians often decorated their tepees with colorful ....

Pocahontas falls in Love

After many months, Pocahontas met John Rolfe. He was a young tobacco farmer. John Rolfe was very kind to her. She loved him and he loved her. They were very happy together.

John Rolfe wanted to marry Pocahontas. But Pocahontas wanted to speak to her father first.

One day Pocahontas and John Rolfe went to visit Powhatan. The great chief was pleased to see his daughter.

"Father, this is John Rolfe. I met him in Jamestown. He is a very kind person. I love him and I'm going to marry him."

"You are a young woman now, Pocahontas. It is time for you to marry the man you love." Powhatan embraced[1] his daughter and John.

"Great chief, I love your daughter and I want to take care of her," said John Rolfe.

1. embraced : 擁抱。

Pocahontas falls in Love

In April 1614, Pocahontas married John Rolfe in the Anglican Church. She was the first American Indian to marry a white man. Pocahontas wore a beautiful white dress. She had flowers in her hair. She and John Rolfe laughed and danced on this special day. All the people in Jamestown celebrated the marriage. There were great festivities [1] with music and dancing. Nantaquas and many other Indians came to celebrate. At the wedding meal [2] there were white men and Indians. They were all good friends. They sat around a long table. There were all types of good food to eat. Pocahontas' marriage brought peace to

1. **festivities** : 慶祝活動。
2. **wedding meal** : 婚宴。

the Virginia Colony. The colony began to grow.

Soon after their marriage, Pocahontas and John went to London. Pocahontas was surprised to see many new things in England. In London she wore beautiful English clothes. Everyone wanted to meet her. She met the most important people of London. She even met King James I! Everyone loved the Indian princess.

In London Pocahontas met John Smith again after many years! This was a wonderful surprise. They talked about their adventures in Jamestown.

Pocahontas and John Rolfe had a son. They were very happy. They called him Thomas. After Thomas' birth, Pocahontas became very ill. She died in England in 1617, at the age of 21. This was a tragic [1] event. Many people mourned [2] her. They remembered the courageous Indian princess.

Young Thomas was educated in Britain by his uncle Henry Rolfe. John Rolfe was heartbroken [3]. He didn't want to live in Britain. He returned to Virginia to grow tobacco. He became an important tobacco farmer. He had a very big tobacco plantation.

When Thomas Rolfe was an adult, he left Britain. He went to America to visit his mother's land. He met his mother's tribe. He liked the New World and remained there. In 1788 the Virginia Colony became a state of the United States.

There are still descendants [4] of Thomas Rolfe, Pocahontas' son, in Virginia today.

1. **tragic** : 悲慘的。
2. **mourned** : 哀悼。
3. **heartbroken** : 極傷心的。
4. **descendants** : 後裔。

UNDERSTANDING THE TEXT

 Read the paragraph below.
Choose the best word (A, B or C) for each space (1-8).

Pocahontas wanted to marry John Rolfe ⁰ ...A...... she loved him. Her
father was happy and ¹ .... April 1614 they were married in
Jamestown. On that day there ² .... a big celebration. Her
marriage ³ .... peace ⁴ .... the white people and the Indians.
Pocahontas and ⁵ .... husband visited England. ⁶ .... liked the
Indian princess. ⁷ .... the birth of her son Thomas, Pocahontas died
and John Rolfe was heartbroken. He returned ⁸ .... Virginia.

0. A because	**B** but	**C** and
1. A in	**B** at	**C** on
2. A be	**B** was	**C** were
3. A brought	**B** made	**C** did
4. A for	**B** between	**C** with
5. A her	**B** his	**C** its
6. A All	**B** Anyone	**C** Everyone
7. A Before	**B** After	**C** During
8. A to	**B** in	**C** at

T: GRADE 3

 Topic – Places/Weather
"Pocahontas was surprised to see many new things in England."

Bring in a photo or picture of something you can see in your town.

   a. Describe the place(s) in your picture.

   b. What's the weather like in the picture?

   c. What other interesting things can a visitor see in your town?

Have fun with this crossword puzzle!

ACROSS

1. Pocahontas' Christian name.
2. Captain Smith established the settlement of ....
3.
4. Chief of the Potomac tribe.

DOWN

5. The Virginia ....
6. Pocahontas' husband.
7. Pocahontas' son.
8.
9.
10. A type of hut.

57

Who did what?

Read the clues and write the name of the person who did the following actions. Some names can be used more than once.

Who...

a. ☐ invited Pocahontas and Chief Japazaws with his wife to the ship?

b. ☐ didn't want to plant crops and work as farmers?

c. ☐ gave Pocahontas English clothes to wear?

d. ☐ saved Captain Smith's life twice?

e. ☐ met Pocahontas in London after many years?

f. ☐ wanted to kill Captain Smith during the night?

g. ☐ was the chief of the Potomac tribe?

h. ☐ did Pocahontas marry?

i. ☐ brought food to the settlers of Jamestown?

j. ☐ captured Pocahontas?

k. ☐ was educated in Britain but went to live in America?

1. Pocahontas
2. John Rolfe
3. women of Jamestown
4. Captain Samuel Argall
5. Chief Powhatan
6. Captain John Smith
7. Chief Japazaws
8. English gentlemen
9. Thomas Rolfe

Look at this sentence from Chapter 5:

"I love him and I'm going to marry him."
(Pocahontas wants to marry John in the future.)

We can use "going to" to talk about future actions.

Complete the sentences with "going to" and the verbs in the box. One is done for you.

> meet   leave   live   eat
> hunt   declare   dance   visit

a. The settlers are .going to eat........... the corn at midday.

b. Chief Powhatan is .... war tomorrow.

c. She is .... John this evening.

d. John Smith is .... Jamestown next month.

e. Tomorrow Pocahontas is .... with the Potomacs.

f. "I am .... in the forest on Monday," said Nantaquas.

g. They are .... London next week.

h. We are .... at the party.

 Read the descriptions. What is the word for each one? The first letter is already there. There is one space for each other letter in the word.

0. He was very, very sad.   h e a r t b r o k e n

1. She was brave.   c _ _ _ _ _ _ _ _ _

2. All the people.   e _ _ _ _ _ _ _

3. A wedding.   m _ _ _ _ _ _ _

4. He grows crops.   f _ _ _ _ _

5. You are not well.   i _ _

59

7 Write the names of the characters under the pictures.
Then describe them using the words in the tepee.
Some words can be used more than once.

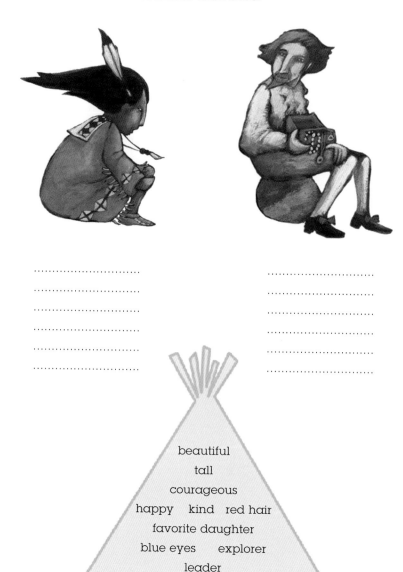

....
....
....
....
....
....

....
....
....
....
....
....

beautiful

tall

courageous

happy   kind   red hair

favorite daughter

blue eyes   explorer

leader

8 Now write a few sentences describing each character.

    a. Pocahontas is ....
....

    b. ....
....

9 What new words did you learn in this story?

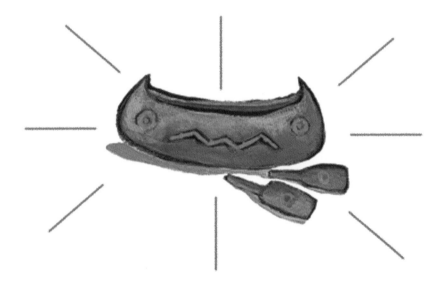

QUESTIONS FOR YOU

    a. Did you see the film *Pocahontas*?

    b. Did you like it?

    c. How is the film different from this story?

    d. Which one do you prefer? Why?

Virginia Today

Today Virginia is an American state. It became a state in 1788, after the American Revolution. About 5 million people live in Virginia. The capital of Virginia is Richmond. Other important cities are Roanoke, Norfolk, Newport and Alexandria.

George Washington was born in Virginia. He is called the Father of America. He was the leader of the American Revolution. He

George Washington (c. 1795) by Rembrandt Peale.

Mount Vernon, George Washington's home.

fought for the independence of the United States. In 1789 George Washington became the first president of the United States.

Today you can visit the Jamestown Settlement. You can see people in 17th-century costumes [1] and how they lived.

Colonial Williamsburg is the biggest 18th-century town in the world!

A woman in a 17th-century costume at the Jamestown settlement.

Many tourists visit the historical sites in Virginia.

1. **costume** : 服裝。

In Virginia today there is the Pocahontas State Park. The beautiful park was named after Pocahontas. There is also a museum dedicated to [1] the Indian princess.

There are many excellent colleges and universities in this beautiful state.

1 **You have the answers. Now ask the questions.**

a. Q: .... Virginia become a state in the 17th century?
A: No, it didn't.

b. Q: .... many people live in Virginia?
A: Five million people live in Virginia.

c. Q: .... Roanoke the capital of Virginia?
A: No, it isn't.

d. Q: .... Washington the first president of the United States?
A: Yes, he was.

e. Q: .... there a state park named after Pocahontas?
A: Yes, there is.

1. **dedicated to** : 紀念。

PROJECT ON THE WEB

LET'S FIND OUT MORE ABOUT POCAHONTAS AND THE OTHER CHARACTERS OF THE STORY!

Find the correct Web site and answer these questions.

a. What was Pocahontas' real name?

b. Did any of the other characters have different names? What were they?

c. How many wives did Chief Powhatan have?

d. What can you find out about John Rolfe?

e. Can you find other information about Pocahontas and her times?

f. What differences are there between the story you read and the information on the Web site?

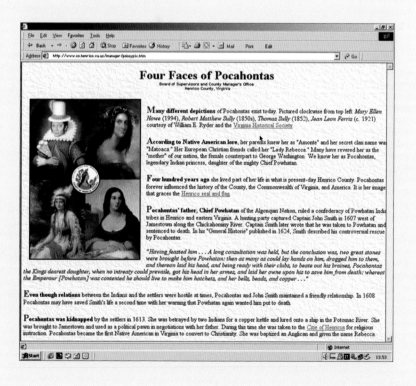

EXIT TEST

 Read the paragraphs below about the story of Pacahontas. Choose the best word (A, B or C) for each space (1-8).

Pocahontas was the ⁰ ..B....... daughter of Chief Powhatan of the Algonquin tribe. When the white men ¹ ...., Chief Powhatan was not happy because ² .... wanted to take the Indians' land.

Pocahontas was a courageous girl and saved Captain John Smith's life ³ .... They became good friends. ⁴ .... the winter, Pocahontas brought food to the settlers in Jamestown. One day Captain Smith returned to England.

After six years Chief Powhatan ⁵ .... war on Jamestown. He sent his daughter to live with the Potomac tribe. Captain Samuel Argall tricked ⁶ .... and she became a prisoner. But Pocahontas liked Jamestown and ⁷ .... people. She married John Rolfe and ⁸ .... to England. She died after the birth of her son Thomas. Many people mourned her.

0. A best	**B** favorite	**C** preferred
1. A happened	**B** arrived	**C** entered
2. A he	**B** them	**C** they
3. A twice	**B** two	**C** double
4. A By	**B** During	**C** At
5. A said	**B** told	**C** declared
6. A her	**B** she	**C** hers
7. A their	**B** its	**C** his
8. A gone	**B** going	**C** went

Score ▭

Complete the conversation.
What does Pocahontas say to Nantaquas?
For questions 1-5, mark the correct letter A-H.

Pocahontas: Can we go to see the sailing ship?
Nantaquas: 0 F..........
Pocahontas: When can we go?
Nantaquas: 1 ....
Pocahontas: Let's take the small canoe.
Nantaquas: 2 ....
Pocahontas: How long will it take to get to the ship?
Nantaquas: 3 ....
Pocahontas: It's a beautiful ship.
Nantaquas: 4 ....
Pocahontas: Do you see a white man smiling at us?
Nantaquas: 5 ....
Pocahontas: He's friendly.

A That's a good idea.
B It really is.
C About thirty minutes.
D Yes, I do.
E No, I won't.
F Yes, of course we can.
G In an hour.
H Where is it?

Score

3 Match the people (1-6) with what they did (A-F).

A ☐ He was an important tobacco farmer.
B ☐ He was Pocahontas' brother.
C ☐ He tricked Pocahontas.
D ☐ He was Pocahontas' father.
E ☐ He was Thomas Rolfe's uncle.
F ☐ He was the chief of the Potomac tribe.

1. Captain Samuel Argall
2. Henry Rolfe
3. Chief Japazaws
4. Nantaquas
5. John Rolfe
6. Chief Powhatan

Score

4 Read the sentences and choose the best word (A, B or C) for each space (1-5).

0. The settlers ...A..... some huts and a storehouse.
   A built   **B** did   **C** made
1. Pocahontas was the .... daughter of Chief Powhatan.
   A better   **B** best   **C** favorite
2. Nantaquas ran to .... his father about the sailing ship.
   A talk   **B** say   **C** tell
3. They got .... a small canoe.
   A by   **B** into   **C** in
4. The Indians .... the white men carefully.
   A watched   **B** saw   **C** looked
5. "You must all .... this land now!" said the chief.
   A away   **B** go   **C** leave

Score

The True Story of Pocahontas

KEY TO THE EXERCISES

BEFORE YOU READ

Page 10 Exercise 2
a. hut or cabin
b. settlers
c. wilderness
d. canoe
e. medicine men
f. farmer

CHAPTER ONE

Page 15 Exercise 1
1. B 2. B 3. C 4. B 5. C 6. A 7. A

Page 15 Exercise 2
tall: tree, man, child, daughter
colourful: dress, tree, flowers
sweet: food, child, biscuit
blue: dress, eyes, sky, flowers
bad: man, food, child, biscuit, weather, daughter
strong: man, child, tree, daughter

Page 16 Exercise 3
1. C 2. A 3. C 4. B 5. C

Page 17 Exercise 4
a. eleven
b. eighteen
c. three hundred
d. eight
e. fourteen
f. thirty-two
g. seventy-five

Page 17 Exercise 5
1. girl 2. happy 3. sea 4. eleven
5. black 6. father 7. brother
8. old 9. like

CHAPTER TWO

Page 24 Exercise 1
a. Jamestown
b. English gentlemen
c. find gold and riches
d. went into the forest to look for food
e. to kill John Smith
f. put her head above Captain Smith's
g. they became great friends

Page 25 Exercise 2
1. storehouse
2. compass
3. chief
4. settlement
5. palace

Page 25 Exercise 3
a. They sit on the grass.
b. The weather is very good.
c. He is showing Pocahontas beartiful beads and trinkets.
d. Open answer.
e. John Smith taught Pocahontas English.

Page 26 Exercise 4
a. 5 b. 1 c. 2 d. 3 e. 4 f. 2
g. 5 h. 4 i. 3 j. 3

Page 26 Exercise 5
Open questions.

Page 27 Exercise 6
1. D 2. G 3. H 4. B 5. A

Page 28 Exercise 7
1. was 2. called 3. not 4. want
5. plant 6. some 7. are 8. us
9. an 10. her

THE ORIGINS OF VIRGINIA

Page 32 Exercise 1
1. A 2. B 3. C 4. B 5. A 6. C 7. B
8. A

Page 32 Exercise 2
a. ELIZABETH
b. SCIMITAR
c. COAT OF ARMS
d. PRISONER
e. GOLD

CHAPTER THREE

Page 35 Exercise 1
1. C 2. A 3. A 4. B 5. A 6. A 7. C
8. B

Page 36 Exercise 2
a. messenger
b. summer
c. basket
d. food
e. winter
f. longhouse

1d – 2c – 3f – 4a – 5e – 6b.

BEFORE YOU READ

Page 36 Exercise 1
a. She is in Jamestown.
b. White women in Jamestown.
c. She is wearing English clothes.
d. She is happy.

Page 37 Exercise 2
1. A 2. C 3. A 4. A 5. B 6. C

CHAPTER FOUR

Page 42 Exercise 1
a. and Chief Powhatan was angry

and declared war.
b. were friends of the white people.
c. to live with the Potomac tribe.
d. an English explorer.
e. to see his sailing ship.
f. helped Captain Argall to capture Pocahontas.
g. and she made many friends and became a Christian.

Page 43 Exercise 2

M	N	B	D	H	S	H	Y	D	O	A	G	Y
G	A	P	J	C	P	O	W	H	A	T	A	N
X	N	L	E	L	W	U	E	Z	S	M	K	L
O	T	V	G	P	H	T	I	W	G	N	J	R
F	A	L	P	O	C	A	H	O	N	T	A	S
S	Q	U	A	T	N	K	C	F	P	H	P	N
T	U	F	I	O	P	Q	E	S	F	S	A	O
W	A	Y	C	M	V	A	U	G	V	Y	Z	V
I	S	Q	K	A	J	C	X	I	K	R	A	E
B	E	A	L	C	B	F	T	P	N	H	W	R
R	O	C	Z	R	D	G	W	F	L	K	S	N

Page 43 Exercise 3
a. Powhatan, Algonquin
b. Japazaws, Potomac
c. Nantaquas and Pocahontas

Page 43 Exercise 4
Open questions.

Page 44 Exercise 5
a. What b. Where c. Why
d. When e. Who

Page 44 Exercise 6
a. safe, tribe
b. tall, sails
c. colored, payment
d. people, kind
e. brave, life
f. dangerous, forest

Page 45 Exercise 7
1. prisoner 2. Jamestown 3. wife
4. true friends 5. kind 6. free
7. weapons 8. settlers 9. me
10. daughter

Page 45 Exercise 8

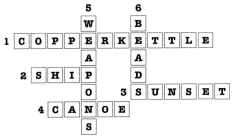

Page 57 Exercise 3

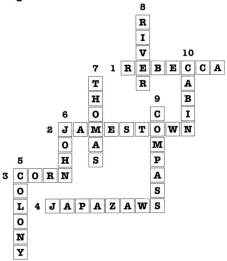

INDIAN LIFE

Page 51 Exercise 1
a. F – The American Indians came from Asia about 40,000 years ago.
b. T
c. F – American Indians lived in tribes.
d. F – The Indians painted pictures on big stone walls.
e. T
f. T
g. T
h. F – The Apaches, the Comanches and the Mohicans were important tribes.
i. T
j. T

Page 51 Exercise 2
a. pictures; b. buffalo; c. tomahawk; d. totem poles; e. nature

Page 51 Exercise 3
a. totem poles   b. buffalo   c. nature
d. tomahawk   e. pictures

CHAPTER FIVE

Page 56 Exercise 1
1. A 2. B 3. A 4. B 5. A 6. C 7. B 8. A

Page 56 Exercise 2
Open questions.

Page 58 Exercise 4
a. 4 b. 8 c. 3 d. 1 e. 6 f. 5 g. 7
h. 2 i. 1 j. 4 k. 9

Page 59 Exercise 5
b. going to declare
c. going to meet
d. going to leave
e. going to live
f. going to hunt
g. going to visit
h. going to dance

Page 59 Exercise 6
1. courageous
2. everyone
3. marriage
4. farmer
5. ill

Page 60 Exercise 7
Pocahontas: beautiful, courageous, happy, kind, favorite daughter
John Smith: tall, courageous, kind, red hair, blue eyes, explorer, leader

Page 61 Exercises 8-9
Open questions.

VIRGINIA TODAY

Page 64 Exercise 1
a. Did **b.** How **c.** Is **d.** Was   e. Is

1. **1.** B **2.** C **3.** A **4.** B **5.** C
    6. A **7.** B **8.** C
2. **1.** G **2.** A **3.** C **4.** B **5.** D
3. **A.** 5 **B.** 4 **C.** 1 **D.** 6 **E.** 2 **F.** 3
4. **1.** C **2.** C **3.** B **4.** A **5.** C

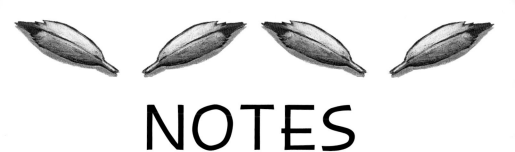

NOTES

NOTES

NOTES

NOTES

NOTES

This Chinese edition of *The True Story of Pocahontas* has been published with the written permission of Black Cat Publishing.

The copyright of this Chinese edition is owned by The Commercial Press (H.K.) Ltd.

Name of Book: The True Story of Pocahontas
Told by: Kelly Reinhart
Editors: Monika Marszewska, Elvira Poggi Repetto
Design and art direction: Nadia Maestri
Computer graphics: Simona Corniola
Illustrations: Barbara Nascimbeni
Edition: ©2003 Black Cat Publishing
            an imprint of Cideb Editrice, Genoa, Canterbury

系 列 名： Black Cat 優質英語階梯閱讀 · Level 1
書　　名：風中奇緣
責任編輯：傅　伊
封面設計：張　毅
出　　版：商務印書館 (香港) 有限公司
　　　　　香港筲箕灣耀興道 3 號東滙廣場 8 樓
　　　　　http://www.commercialpress.com.hk
發　　行：香港聯合書刊物流有限公司
　　　　　香港新界大埔汀麗路 36 號中華商務印刷大廈 3 字樓
印　　刷：中華商務彩色印刷有限公司
　　　　　香港新界大埔汀麗路 36 號中華商務印刷大廈
版　　次：2015 年 3 月第 8 次印刷
　　　　　©2003 商務印書館 (香港) 有限公司
　　　　　ISBN 978 962 07 1637 9
　　　　　Printed in Hong Kong

Black Cat English Readers

BLACK CAT ENGLISH CLUB
Membership Application Form

BLACK CAT ENGLISH CLUB is for those who love English reading and seek for better English to share and learn with fun together.

Benefits offered: - *Membership Card*

- *Member badge, poster, bookmark*

- *Book discount coupon*

- *Black Cat English Reward Scheme*

- *English learning e-forum*

- *Surprise gift and more...*

Simply fill out the application form below and fax it back to **2565 1113**.

Join Now! It's FREE exclusively for readers who have purchased *Black Cat English Readers* !

The book(or book set) that you have purchased: _____

English Name:_____ (Surname) _____ (Given Name)

Chinese Name: _____

Address:_____

Tel: _____ Fax: _____

Email:_____
(Login password for e-forum will be sent to this email address.)

Sex: ❏ Male   ❏ Female

Education Background: ❏ Primary 1-3   ❏ Primary 4-6   ❏ Junior Secondary Education (F1-3)

❏ Senior Secondary Education (F4-5)   ❏ Matriculation

❏ College   ❏ University or above

Age: ❏ 6 - 9   ❏ 10 - 12   ❏ 13 - 15   ❏ 16 - 18   ❏ 19 - 24   ❏ 25 - 34

❏ 35 - 44   ❏ 45 - 54   ❏ 55 or above

Occupation: ❏ Student   ❏ Teacher   ❏ White Collar   ❏ Blue Collar

❏ Professional   ❏ Manager   ❏ Business Owner   ❏ Housewife

❏ Others (please specify: _____)

As a member, what would you like **BLACK CAT ENGLISH CLUB** to offer:

❏ Member gathering/ party   ❏ English class with native teacher   ❏ English competition

❏ Newsletter   ❏ Online sharing   ❏ Book fair

❏ Book discount   ❏ Others (please specify: _____)

Other suggestions to **BLACK CAT ENGLISH CLUB**:

_____

Please sign here: _____

(Date:_____)